HYBRID A PROJECT MANAGEMENT

HYBRID MODEL COMBINING SCRUM, KANBAN AND SOME WATERFALL TECHNIQUES TO DRIVE INNOVATION AND OPERATIONAL EXCELLENCE

BUILDING STATE OF THE ART CUTTING EDGE PRODUCTS WITH PASSION, QUALITY AND SPEED

VIJAYAKRISHNA R. N.

Copyright © 2020. All Right Reserved.

No part of this publication may be reproduced, distributed, or transmitted in any form or by any means, including photocopying, recording, or other electronic or mechanical methods, or by any information storage and retrieval system without the prior written permission of the publisher, except in the case of very brief quotations embodied in critical reviews and certain other noncommercial uses permitted by copyright law.

Table of Contents

Intent 7
Why You Should Read This Book 8
Part 1: Agile Methodologies Overview and Benefits 10
    1.1 Agile Methodologies Overview 10
        Background & Core Values 10
    1.2 List of Agile Methodologies 11
    1.3 Agile Methodology Benefits 12
Part 2: Agile Problems 13
    2.1 The "Compromised" Architecture 13
        Compromised Architecture 13
        High Maintenance (Death by thousand cuts) 13
        Poor Product Quality (Quick and dirty way) 13
    2.2 The "Scrum Master" Problem 14
        Giving control to non-technical "Scrum Master"! 14
        Product Releases that often Slip! 14
        Process Misunderstood 14
    2.3 The "Demotivated" Employees 15
    2.4 The "Creativity" that's Killed 16
    2.5 The "Tactical" Product Vision 16
Part 3: Introduction to **Hybrid Agile** 17
    3.1 Hybrid Agile Overview 17
        Adapting Scrum/Kanban to make it Work with some Waterfall 17
        Why Time-Boxing is Important 18

    3.2 Hybrid Agile Goals..19

    3.3 Hybrid Agile v/s Other 21

Part 4: **Hybrid Agile** Project Lifecycle 24

    4.1 Lifecycle Overview..24

    4.2 The RACI Matrix...27

Part 5: **Hybrid Agile** Process Guidelines 31

    5.1 Product Backlog 31

        Product Backlog..31

        User Stories...32

        Where do Stories come from? 32

        The Backlog Clutter..33

        Good Story Definition..36

    5.2 Release Backlog and Scoping 38

        The Ordered Flat List..38

        Estimation Process..40

        Person Days, Fibonacci and T-Shirt methods 41

        Scope To Release 44

        Uniform Story Points 46

        Estimation Tool...47

    5.3 Design Sprint...48

        Analysis 48

        Solution Architecture 48

        Prototyping using Mock-ups..49

        CI/CD and Source Code 49

        Source Code and Branching Strategy 50

- CI/CD 50
- Test Plan 51
- 5.4 Dev Sprints and Beta Releases 52
  - Sprint 52
  - Sprint Planning 52
  - Stories and Sub-tasks 56
  - Standups 57
  - Key Metrics Logging 58
  - Design & Development 59
  - Error Handling and Error Codes 60
  - Automation Test Suites 62
  - Test Scripts 64
  - Story Workflow 64
  - Defect Workflow 65
  - API & User Documentation 66
  - Sprint Demo 67
  - Feedback Loop 67
  - Splitting Stories 68
  - Sprint Closing 69
  - Beta Release 69
- 5.5 Release Candidate Sprint 69
  - Release Candidate 69
  - Non-Functional Testing 70
- 5.6 Release To Production (GA Release) 70
  - Checklist 70

    Deployment 71

Part 6: **Hybrid Agile** Project Reports 72

    6.1 Reports Overview 72

    6.2 Release Report 73

    6.3 Sprint Report 74

    6.4 Velocity Report 76

    6.5 Time Log Report 77

    6.6 Product Backlog Report 78

Part 7: **Hybrid Agile** Areas of Excellence Handbook 79

    7.1 Excellence in Product Strategy 79

    7.2 Excellence in Customer View 80

    7.3 Excellence in Architecture 81

    7.4 Excellence in Quality 81

    7.5 Excellence in Innovation 82

    7.6 Excellence in Automation 83

    7.7 Excellence in Speed-To-Market 84

Part 8: Scaled Agile Considerations 86

About The Author 89

References 90

Disclaimer 90

One Last Thing 90

Intent

According to the U.S. Bureau of Labor Statistics [1] (BLS) approximately 20% of new businesses fail during the first two years of being open, 45% during the first five years, and 65% during the first 10 years!

Agile **methodology** greatly helps companies to build products in iterations, get timely feedback, achieve speed to market, capture market share early, fail fast, adapt to changes quickly and more efficiently as they grow and aim to succeed.

There are various types of agile methodologies in practice – Scrum, Kanban, FDD, and others.

The Scrum is one of the most popular agile methodology in use by companies.

While Scrum can be successfully implemented to execute the project, Scrum's implementations can also cause several problems in a complex product type of application where simply executing each sprint as per strict agile Scrum doctrine is not practical.

Some of these problems are – compromised architecture, challenges in handling dependencies, demotivated employees due to routine and time pressures, lack of creativity and innovation culture, difficulty closing the story within the sprint, too much tactical product vision instead of strategic, development environment that's like stress chamber etc.,

In author's experience considerable amount of success can be achieved by taking Scrum as base process and then blending some Kanban and Waterfall methodologies resulting in Hybrid Agile project management process.

The primary goal of this book is to outline this Hybrid Agile process for such product development projects where while pure agile concepts like Scrum time-boxing is used, some non-agile concepts are also used to overcome the drawbacks of pure agile implementations.

Hybrid Agile is not a defined framework like Scrum or Kanban. The specification for Hybrid Agile given in this book is author's definition and implementation of same by using maximum Scrum and customizing and optimizing with some Kanban, Waterfall etc.

WHY YOU SHOULD READ THIS BOOK

This book can greatly benefit if:

- As an organization you are evaluating agile methodologies and would like to learn issues faced in agile methodologies especially for products that takes more than one sprint to release to customer.

- As an organization you are currently using agile methodologies and you see opportunities to further innovate and achieve operational excellence.

- As an agile project lead you have been using Scrum/Kanban methodologies but you are seeing the architecture is compromised to an extent and there is lack of innovation happening.

- As an engineering/people manager you are facing issues around employee happiness, engagement levels, creativity and general motivation and learn how to address them.

This book is not intended to be:

- This book is not intended to be a pure Scrum guideline book or Kanban guideline as the guide is intended to customize Scrum, Kanban and also uses some non-agile concepts to better achieve project delivery with innovation and operational excellence.
- However it may still be useful for pure Scrum practitioners to learn about various concepts such as estimation, planning etc.

PART 1: AGILE METHODOLOGIES OVERVIEW AND BENEFITS

1.1 AGILE METHODOLOGIES OVERVIEW

Agile methodology is a blanket term used for various software development methodologies that have been influenced by agile core principles *of iterative and incremental development process* to build software products.

BACKGROUND & CORE VALUES

Agile development model has been applied for years in various forms without a formal definition for years. In 2001 a group of developers released an Agile Manifesto in Snowbird summit that has since influenced several methodologies being developed inspired by core values.

The **Agile** Manifesto [3] is comprised of 4 foundational **values and** 12 supporting **principles** which lead the **agile** approach to software development.

The *Manifesto for Agile Software Development* **Values**

- ✓ **Individuals and interactions** over processes and tools.
- ✓ **Working software** over comprehensive documentation.
- ✓ **Customer collaboration** over contract negotiation.
- ✓ **Responding to change** over following a plan.

The *Manifesto for Agile Software Development* principles

1. Customer satisfaction by early and continuous delivery of valuable software.
2. Welcome changing requirements, even in late development.
3. Deliver working software frequently (weeks rather than months).
4. Close, daily cooperation between business people and developers.
5. Projects are built around motivated individuals, who should be trusted.
6. Face-to-face conversation is the best form of communication (co-location).
7. Working software is the primary measure of progress.
8. Sustainable development, able to maintain a constant pace.
9. Continuous attention to technical excellence and good design.
10. Simplicity—the art of maximizing the amount of work not done—is essential.
11. Best architectures, requirements, and designs emerge from self-organizing teams.
12. Regularly, the team reflects on how to become more effective, and adjusts accordingly.

1.2 List of Agile Methodologies

There are several agile methodologies inspired by the Agile Manifesto core values and principles – The most popular are listed below:

- Agile Scrum Methodology
- Lean Software Development
- Kanban
- Extreme Programming (XP)

- Crystal
- Dynamic Systems Development Method (DSDM)
- Feature Driven Development (FDD)

More comprehensive list of agile methodologies can be found in agile software development wiki. [4]

1.3 Agile Methodology Benefits

There are several benefits to adopting agile methodologies – key ones are as follows:

1. **Initial Business Value** – this is one of the major benefits of agile - Through the process of continuous incremental delivery and continuous feedback cycle, agile helps align delivered software with desired business needs.
2. **Adapting to Change** – with various micro and macro factors impacting the business, adapting to change becomes a very key factor in success or failure of a product. Through incremental delivery mechanism adapting to change is better managed with agile.
3. **Improved Customer Engagement** – with frequent releases, the customer engagement improves with agile.

There are other benefits such as increased Productivity and increased Quality if implemented in right methods.

Part 2: Agile Problems

2.1 The "Compromised" Architecture

Compromised Architecture

Following are some of the reasons why product often ends up with compromised architecture.

- Getting ahead of competitor is highest priority.
- Too much business driven without technical leadership participation.
- Lack of product vision and roadmap with belief always can adapt later.

High Maintenance (Death by thousand cuts)

With compromised architecture, product will suffer from high maintenance costs every time there is a change due to fact that change is now extremely difficult to do without impacting various parts of the system. This can quickly choke the product development and eventually lead to situation such as death by thousand cuts.

Poor Product Quality (Quick and dirty way)

The product often doesn't get tested thoroughly due to continuous timeline pressures and due to wrong way of operating sprints. This leads to product getting shipped with defects.

2.2 The "Scrum Master" Problem

Giving control to non-technical "Scrum Master"!

In Waterfall world, product development was primarily driven by engineering leadership. This is primarily driven by engineering manager or technical development lead with team of architects, developers, test and infra teams.

However with agile development – Scrum introduces a new role "Scrum master" to run agile development team.

Often organizations make a mistake of hiring Scrum masters who may sometimes lack in-depth knowledge and technical capabilities of product development.

If you provide control of product development to Scrum masters who are not experienced in technical aspects of how to build a software product, it may lead to chaos, frustrations among technical team, high stress environment and process for the sake of process with rationale in process missing many times.

Product Releases that often Slip!

Often times, work breakdown structure is not fully visualized by the Scrum masters and developers. This results in additional key stories being discovered as the project progresses which leads to project releases slipping.

Process Misunderstood

Process is important for any organization but it needs to be optimized and automated to maximum level to reap its full benefits. Often times Scrum masters lack the maturity to optimize the process – this results in process being inhibitor than catalyst.

Some examples of these are such as estimation poker kind of activity done with more process in mind, less technical expertise leading to incorrect estimates – this is explained in later parts of the book.

2.3 The "Demotivated" Employees

If agile, especially agile Scrum methodologies are not implemented properly, it often leads to demotivated employees.

Few primary reasons for demotivation in workplace are:

- **Micromanagement** -- the strict guidelines around daily standups, etc., can easily move management styles towards micromanagement.
- **Short term objectives with no career vision** - agile assumes the entire team is at same skill level, and doesn't address how performance should be measured in agile team as everything is team success or team failure.
- **Boredom with routine** – with time boxing and frequent releases approach, the project development can become extremely routine and cause boredom.
- **Stressful environment** – with frequent deliveries on weekly/bi-weekly sprints – the environment can be very stressful to development teams if not work is well organized. Development environment may feel like stress chamber.

Other reasons:

- **Poor leadership** – this is another major reason for employee demotivation however this cannot be attributed to agile methodologies alone.

2.4 The "Creativity" that's Killed

Developers who are otherwise extremely creative and innovative are trapped in a system, where everything is like a manufacturing factory line, everything routine and stressful.

Developers quickly lose focus on creativity and innovation as they are always caught up in heavy duty Scrum process – planning, estimation, daily standup updates, story completion, and other -- all during a short duration and that may feel like very highly stressful environment.

2.5 The "Tactical" Product Vision

Great product companies typically have following:

- A clear product vision,
- A solid product strategy,
- A set of priorities and
- A way to measure outcomes.

With agile development, often mindset is set in product management leadership that they can always adapt at great speed, and therefore there is no need to think beyond six months for example. This may lead to crappy product that can eventually get stagnant.

Part 3: Introduction to **Hybrid Agile**

3.1 Hybrid Agile Overview

The primary goal of "Hybrid Agile" guidelines is to look at pure agile methodology, mainly Scrum methodology, look at problems in implementing Scrum, and provide guidelines to customize and optimize the process to help achieve innovation and operational excellence and to help build state of the art cutting edge products.

In this process of customizing, some non-agile concepts are included such as separate QA process, non-functional requirement QA execution and other.

In nutshell, the Hybrid Agile goal is to maximum adhere to agile guideline but at times take liberty to use non-agile concepts as well to achieve project delivery with innovation and operational excellence.

Adapting Scrum/Kanban to make it Work with some Waterfall

Scrum Methodology [5] is one the most popular agile framework and organizations can adapt this methodology and succeed.

In the next parts of this book, we will look into various aspects of Scrum and how to customize and optimize better for product development.

Hybrid Agile guidelines are arrived at by primarily practicing Scrum for long time for product development (especially quarterly development cycles) and then looking at what necessary customizations and optimizations can be made to drive innovation and operational excellence and goals as outlined in section 3.2.

With these customizations and optimizations made, by using mix of agile time-boxing concept from Scrum, flat list concept from Kanban and separate QA process for system testing from Waterfall and few other, the concepts has been quite successful in author's experience for such a use case.

WHY TIME-BOXING IS IMPORTANT

Scrum introduces concept called "time-boxing", also called sprint. Time-boxing is an extremely important aspect of project management.

1. **It allows setting of goals relative to time**. Goals are the oxygen to dreams in every aspect of life and Project management is no different.
2. **Enables relative positioning of work**. Working hard and trying your best sometimes not actually what's required. The alternative – getting the right thing done at the right time – is a better outcome for all. [10]
3. **Enables communicate and collaborate more effectively**. Every member of team can check what is being worked on – since the list is small and well published, it leads to improved collaboration.
4. **Gives you better focus.** Let us say you have just started playing tennis, and your goal is want to improve. This is too generic goal – now let us revise this to say "you want to improve in your forehand in 1 week" – This will keep you focused on watching only forehand training videos,

practicing more forehand, keeps you razor focused on this goal yielding in better results.
5. **Allows you to measure progress easily.** It gives you comprehensive record of what you are doing and what you have done. It helps both individual as well as team.
6. **Gives you motivation.** Time-boxing allows you to aim for concrete end point and get excited about it. Whether tennis, or cycling or running marathon or software development – aiming concrete end points with time-boxing helps you excited and motivated to aim to achieve the goals.

3.2 Hybrid Agile Goals

The primary mission of this book is to provide guidelines to develop product/application using agile methodology, specifically by providing guidelines to customize and optimize Agile Scrum Methodology with some Kanban and including some Waterfall.

With this overall mission, the books aims to address achieving excellence in following areas as well that otherwise might lack in agile environment --

1. **Excellence in Product Vision:** To optimize the processes to drive strategic product vision.

2. **Excellence in Customer View:** To optimize the processes to keep customer in mind all times while building the product.

3. **Excellence in Architecture:** To optimize the processes to drive architecture excellence – product should be build solid architecture principles and best practices.

4. **Excellence in Quality:** To optimize the processes to drive quality – product should be built with high quality.

5. **Excellence in Innovation:** To optimize the processes to drive innovation – product development should have innovation culture cultivated and demonstrated.

6. **Excellence in Automation:** To optimize the processes to drive automation efficiency – processes should be fully automated to maximum possibility.

7. **Excellence in Speed-To-Market:** To optimize the processes to drive speed to market – Processes should enable achieving quickly building and deploying products.

Note: The words "Product" and "Software Application" that is being built are interchangeably used in the book.

3.3 Hybrid Agile v/s Other

The following diagram illustrates how Hybrid Agile is compared to pure Scrum, Kanban and Waterfall.

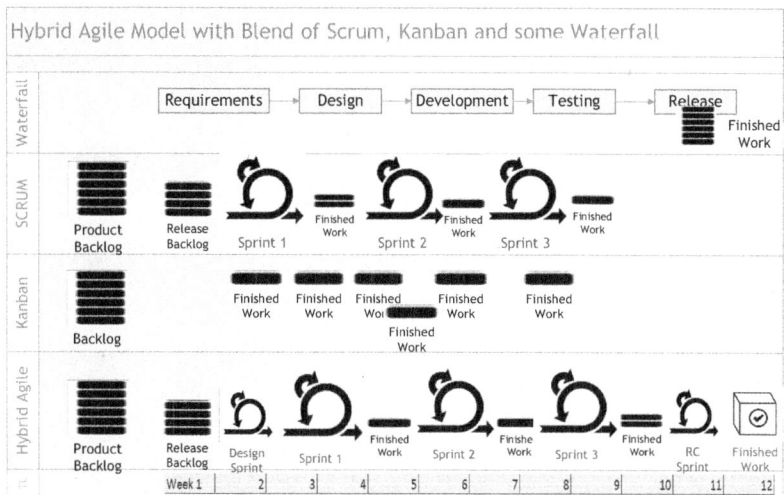

The above diagram illustrates a comparable project lifecycle for a project that has six features written as requirements OR user stories as represented in rectangle black box, for approx. 12 week period.

- **Waterfall**: In Waterfall methodology all the requirements are worked together as a set of requirements and the life cycle involves requirements, design, development, testing and release phase. The finished work is delivered to the customer only at the end of project life cycle (in the above example at the end of 12 week period) and therefore, this poses huge risks in terms of changing market dynamics, changing customer needs, competitors occupying market share during this time, design issues being discovered too late, gap in what product management expected v/s what was delivered and so on.

- **Scrum:** In pure Agile Scrum methodology, the features are categorized as user stories and worked in time-boxes called sprints. Each sprint is planned by estimating user stories, continuously monitoring, designing, developing, testing and delivering – all within the sprint. In the above example as illustrated some amount of finished work is happening every sprint.

- **Kanban:** Kanban is similar to Scrum except that there is no time-box concept and there is no overhead associated with it such as planning, estimation, daily standups, and other sprint related activities. This works best in a more predictable nature of work such as maintenance projects, stories that have very similar characteristics. Due to this simplicity the productivity in Kanban tends to be higher compared to Scrum.

- **Hybrid Agile:** Hybrid Agile is basically taking Scrum as base guidelines and mixing bit of Kanban and Waterfall into it. As illustrated in the diagram above, the sprints are very similar to Scrum except that there is little bit of additional process upfront with the introduction of design sprint and an additional sprint added to the end. Design Sprint is usually very short, **One** week length and therefore is not same length as the rest of sprints. Release Candidate Sprint also may have different length or same as development sprints.

Also, the sprint guidelines need not be very strict as pure Scrum Sprints, there is some flexibility built to it as described in this book.

There is no defined specification as such for Hybrid Agile, therefore the specification for Hybrid Agile given in this book is author's definition and implementation of same.

PART 4: **HYBRID AGILE** PROJECT LIFECYCLE

4.1 LIFECYCLE OVERVIEW

Following is the Hybrid Agile development lifecycle diagram using Scrum methodology and customizing and optimizing using Kanban and Waterfall.

The number of sprints in a release cycle varies product to product, based on type of product. A cloud portal may have lesser sprint cycles compared to a downloadable software product.

However the below overall project lifecycle typically stays the same with variations in length of sprint, number of sprints and need for optional release candidate (system test) sprint.

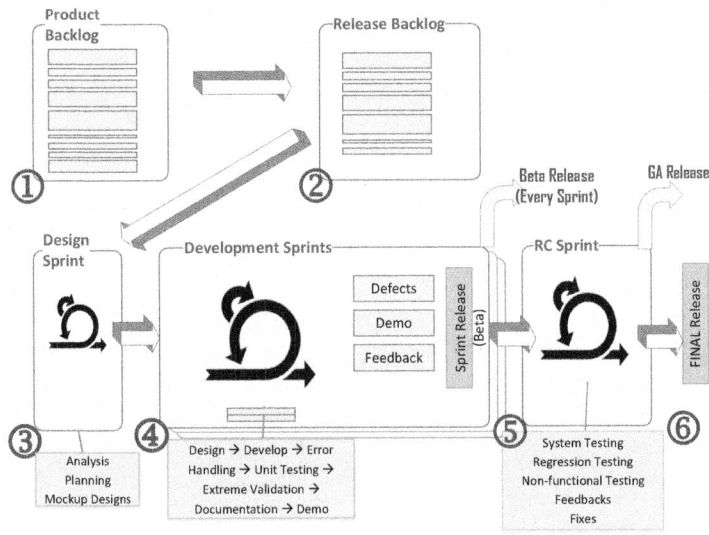

The lifecycle consists of **6 major steps.** Each of these steps illustrated in diagram are explained below:

1. **Product Backlog** is a flat list of all the desired features of the product. Each feature or set of features to be implemented is written as a "User Story". Product backlog therefore is a flat list of all open User Stories.

2. **Release Backlog** is a subset of Product Backlog that is ordered by priority that are identified to be taken up/worked upon in the release. How many stories to be taken up by a release is determined by calculating project team velocity per sprint and how many sprints are available in the release. These concepts are explained in later section of the book.

3. **Design Sprint:** A short one week design sprint is extremely necessary and useful. In these 5 days, the release backlog stories are thoroughly analyzed, solution architecture is put together at a high level and prototyping using mock ups are prepared and validated with product management.

   This ensures development sprints can better focus on individual story design and development since both solution architecture and mock up designs are already complete in design sprint.

4. **Development Sprints**: Stories from the release backlog are taken up for development in a "time-boxed" time frame called Sprint. All development sprints have same length (usually 1 to 4 weeks each)

One or more stories are taken up in sprint. Each story goes thru design, development, unit testing, extreme validation, error handling, documentation, demo and acceptance.

The exact number of development sprints per release depend from product to product.

At the end of sprint, a beta release of the product can be made depending on type of product and organization goals.

Any story that is not completed within same sprint is moved to next sprint.

Defects – As the story gets accepted, any issue that is NOT closed within the sprint cycle is logged as a defect which are taken up to fix by developers as part of regular sprint cycle. However, defects are NOT estimated.

Feedbacks – As the story gets demonstrated and accepted, any feedback from stakeholders/customers are logged as special type of story "Feedback Story". This now becomes a part of release backlog as yet another item and goes thru same process of estimation and being taken up in sprint. These stories are usually small and accommodated within the same release cycle.

5. **Release Candidate Sprint –** Once all the development sprints are complete and stories are accepted, there is usually ONE release candidate sprint that may vary in length compared to development sprint length based on complexity of project.

During this sprint primarily product system testing, regression testing and various non-functional testing such as Scale, Load, Perf, CHO, Browser Compatibility, Device Compatibility etc tests are performed.

What kind of tests are applicable as a *Release Candidate Testing Strategy* depends on the type of product and complexity.

Any critical/high severity issues found are fixed in this sprint lifecycle. The change is typically minimized in this cycle to only must-fix defects.

6. **Product Release:** Product is now fully developed and tested and ready for release. Product is released as a beta-strategy and/or full release.

Each of these steps are explained in detail in Part 5 of this book.

4.2 THE RACI MATRIX

Following are typical roles and people involved in an Agile Project.

Role Key	Role Name	Role Details	Who can play this role?
PO	Product Owner	Product Manager or Product Owner defines user stories.	Product Manager for majority stories. For some stories Engineering manager.
SM	Scrum Master	Scrum Master manages agile	Engineering Manager.

		project execution.	
PTL	Project Technical Lead	Project Technical Lead is the lead responsible for development of product.	Either Engineering Manager OR Technical Architect.
D	Developer	Developers of the product.	Developers.
QAE	QA Engineer	Testers of the product.	QA Engineers.
BE	Build Engineer	Responsible for CI/CD process.	Build Engineer.
RE	Release Engineer	Responsible for actual release to production.	Can be same as Build Engineer or different.
TSM	Tech Sales & Marketing	Responsible for sales and marketing campaign.	Technical Sales & Marketing Resource.

- Sometimes a role can be played by multiple people. For example, for some stories PO can be Product Manager and for some PO can be engineering manager. However, same story should not have multiple PO.
- Sometimes two roles can be played by single persons. Especially Scrum master role played by Project Technical Lead as well may yield better results.

A good project execution needs, clear roles and responsibilities defined. This sounds simple and sometimes perceived as given, but when clarity is provided, exceptional results can be achieved.

RACI matrix [9] is a popular way to have clear roles and responsibilities defined.

A- Accountable (Held accountable for the step)

R- Responsible (Primarily Responsible for the step)
C- Contributor (Contributes to step)
I- Informed (kept in loop)

There can be many roles supporting C and I but usually A, R are limited to only one role to drive better accountability and ownership.

Phase/ Step	PO	SM	PTL	D	QA	BE	RE	TSM
Planning									
Writing Stories	A,R	C	C	I	I	I	I	C
Prioritizing	A,R	C	C	I	I	I	I	C
Estimating Stories	I	I	A,R	C	C	I	I	I
Product Foundation									
Skeleton App	I	I	A, R	C	C	I	I	I
Source Code Mgmt	I	I	A	C	C	R	C	I
CI/CD Setup	I	I	A	C	C	R	C	I
Development of Each Story									
Story Design/Development	I	I	A	R	C	I	I	I
Story Unit Testing	I	I	A	R	C	I	I	I
Story Extreme Validation	C	I	A	C	R	I	I	I
Story Demo	C	C	A	R	C	I	I	I
Story Acceptance	A,R	I	C	I	I	I	I	C
Release/Sprint Management									
Release/Sprint Planning	I	A,R	C	C	C	I	I	I
Resolving Issues/Blockers	I	A,R	C	C	C	I	I	I

Release/Sprint Closing	I	A,R	C	C	C	I	I	I
Release Candidate and Release									
System/NFR Test Cycle	I	I	A	C	R	I	I	I
Beta Program	A	C	C	I	I	I	R	C
Release	A	C	C	I	I	I	R	C
Sales & Marketing Campaign	C	C	C	I	I	I	I	A,R

PART 5: **HYBRID AGILE** PROCESS GUIDELINES

5.1 PRODUCT BACKLOG

PRODUCT BACKLOG

Backlog is simply flat list of product requirements that is written from user perspective. User could an end user of the application OR another system consuming the API interface.

Each requirement is represented as a user story.

Good backlog is critical component of agile project management. For any product, having good backlog is essential to planning market strategy, project prioritization and resource planning.

The backlog is ranked by its priority by product manager who works with all the stakeholders. Top backlog items need to be estimated at high level to ensure a good backlog depth is available at all times.

Product backlog consists of all stories that are not completed yet. A subset of this product backlog is taken up in a release to be worked on and this subset is called release backlog.

Story 1
Story 2
Story 3
Story 4
Story 5
Story 6
Story 7
Story 8
Story 9
Story 10
Story 11
Story 12

Product Backlog

	SP	Total
Story 1	13	
Story 2	21	
Story 3	21	
Story 4	13	
Story 5	13	81
Story 6	13	
Story 7	8	21

Release Backlog

USER STORIES

- A feature is group of requirements that is consumed by end users of the system.

- In agile backlog context, a feature is simply a user story.

- A user story is an independent, testable, demo-able (demonstrable) and consumable feature of the product. Once the story satisfies above characteristics, it should be divided to as small as possible.

Independent
Testable
Demo-able
Consumable

Note: The term "user story" and "story" are used interchangeably thru out this book.

WHERE DO STORIES COME FROM?

Stories for a product can come from various sources.

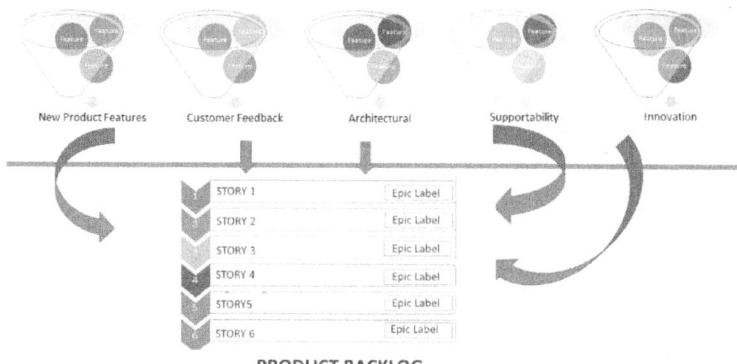

These stories can be organized as given above into various source buckets to help prioritize better:

- Product manager who is accountable for product definition is the primary source of stories. This could include competitive study in the market.
- Customer engagement and customer feedback is another great source for stories.
- Engineering architectural inputs are necessary and drives some stories.
- Various supportability aspects of the products drives some stories.
- Innovation culture in the organization also drives stories and greatly determines the success of the product.

THE BACKLOG CLUTTER

Backlogs can quickly become clutter as product managers and other project team leads add user stories to the backlog.

One practice of backlog structure is, features are organized in hierarchical structure – "Epic" and "User Story". Epic is like a parent feature and User Story is child story of Epic.

Is "Epic" a "Backlog Item" like Story?

As described in previous section backlog is simply flat list of all the user stories.

Epic concept is needed to keep the related stories under a parent feature. This helps product owners to articulate overall vision of the bigger feature being developed so that development team can better design for future.

As explained in the documentation section, the documentation itself is recommended to be in wiki outside of Agile Workflow Tools. Therefore, the Epic to Story relationship should also be maintained in documentation tool.

However, Epic should not be entered or treated as a "backlog item" by itself. That is, it should NOT have its own workflow/lifecycle like story.

Epic should NOT have its own estimation, status, acceptance criteria or any other story attributes as that will only cause clutter and confusion, and breaks away from agile concept of "delivering stories".

- With such an approach treating Epic as a backlog item, now agile team is forced with confusion around should they estimate, test, write test scripts for epic etc.

- It can also make project bit more like Waterfall than agile as multiple stories in epic would mean the progress discussions are always in the context of epic which could be over multiple sprints. It will end up encouraging writing stories that do not adhere to independent, testable, demo-able and consumable principles.
- One of the key to simpler process is to have an ordered **Flat List** of user stories by priority. This way, individuals/team simply picks one after the other, works on it and completes it.
- Hierarchical backlogs breaks the flat list concept to keep things simple. It becomes difficult to visualize priority of stories compared to other. For example priority of Story SU01 under EPIC-A compared to Story SU05 of EPIC-B. This causes quite a bit of manual overhead in the entire process.

So then how to organize Epic in Agile Tools?

Here are couple of options to organize Epic in Agile Tools-

1. Epic can be a parent story provided it doesn't have its own workflow. This means Epic is a parent story for all the child stories, however it will not have workflow attributes such as Estimation, Status, Test Scripts, Acceptance Criteria etc.

2. OR Story attributes can also be used to group the Story to parent-epic (any available attribute or custom attribute). The diagram illustrated below shows how the story attribute is used to label the epic.

Good Story Definition

Bill Wake coins the term INVEST [6] which is one great way to measure if story meets the good definition criteria. INVEST stands for Independent, Negotiable, Valuable, Estimable, Small and Testable.

Considering Negotiable, Valuable, Small are kind of given to agile product development culture - following is another way of measuring if story meets good definition criteria.

Independent	A story should be self-contained so there is no inherent dependency on another story. A story need not be consumed by end user only. It could be another system consuming story. For example in popular micro-service architecture – if there are two services called in end user workflow, they are still considered independent as each service can be independently developed and delivered.
Testable	Extreme Validation is one of the core principles of agile development. Story should provide necessary information to make it fully testable.
Demo-able	Product Management, Team and all Stakeholders need to see what is developed. When story is discussed in planning, it is important to verify what will be demoed at the end of story. If there is no clarity and the answer is ambiguous that means some more work needs to be done to make the story good.
Consumable	Story has been developed, tested and demoed. Who actually consumes it, how it

fits into the larger context of product. Is it another system? Is it end user? This is another checklist to verify if story is defined well.

Should the story title have format "As a <user>"?

There are various flavors of User Story templates. The popular "As a User" [11] template gives clear structure who is the end user and what is the capability.

- **As a <role> I can <capability>,** so that <receive benefit>

As long as these points are captured, the template can be customized but once established first time within the project, it should be kept standard for all stories.

It is important to keep the User Story title short, and therefore the optional "so that" part needs to be excluded from the title itself and put into the Story body instead.

Where to document the story details?

It is a popular practice to use the Agile Tool itself to document the story details, story acceptance criteria and various other attributes of story. **However, this is highly discouraged.** Agile Tools are time-bound workflow tools. They exist to achieve an workflow for the duration of that workflow – once the work flow ends – that is, story is accepted and closed – usually one would never go back and look at the story in agile tool and nor it is easy to find such information in agile tools.

Agile Tools are not made for documentation.

For this reason, documentation about the story should be kept outside of Agile Tools, such as maintain it in wiki software, and should simply linked from Agile Tool for reference.

5.2 Release Backlog and Scoping

The Ordered Flat List

Release Backlog is subset of product backlog that can be scoped to the release.

First step of release scoping is to ensure you have an ordered flat list of stories by priority. This exercise should be done well before release kick off.

Following are some of the key attributes of flat list.

- Story Title – A good title describing the story.
- Label OR a custom field is used to group things together by module/OR higher level functionality.
- Component – indicates category of story, from which bucket the story initiated. Story Bucket or Story Source is a way to organize stories based on where they are originating from – for example – competitive study, innovation, customer feedback, product management, engineering architecture, etc.,
- ROI – product managers can put return on investment so ranking of stories becomes easier.
- Priority – priority of the story in the order it should be picked up for release.

Here is an example backlog to demonstrate priority ranking. In this sample backlog, the application is an invoice management app. Product management prefers to allow registration to app thru self-registration using email but also allows registration using github, facebook, twitter, google login. Product Management believes this will provide more convenience to users in registering to the app. However focus of product management is to first go out with product with minimal functionality to capture market share and then later expand on it, hence #SU04 is ranked below all other.

Story#	Title	Label	Story points	Priority #
SU01	As an unregistered user, I should be able to register using email ID as username and password	Register	13	1
SU02	As a self-registered user I should receive an activation email after registering	Register	13	2
SU03	As a self-registered user my account should be activated in the system upon clicking on activation link	Register	8	3
SU04	As an unregistered user, I should be able to register by logging into my github user id and password	Register	21	7
SU05	As a registered user should be able to login with username/password	Login	13	4
SU06	As a logged in user, I should see dashboard with income and expense reports	Dashboard	21	5

| SU07 | As a logged in user, I should see income/expense be presented as a bar chart per month | Dashboard | 5 | 6 |

ESTIMATION PROCESS

Story Points

Story points are metric used in agile projects to measure how much effort it takes to complete a story.

Estimation process should factor in risks, complexity, and repetitive nature of the problem if problem is already solved. This is not unique to story point method or agile project management method alone to factor this in.

Should you factor in QA effort during Story Point estimation?

Some Agile teams have separate dedicated QA. These QA resources are primarily responsible for system testing, and other non-functional test activities. Unit and functional testing is typically taken care by developer themselves.

Should this QA effort be factored into story estimation? QA effort is traditionally relative to development. It is best to keep things simple and only estimate dev effort in story point estimation process and keep QA bandwidth relative to dev.

This also helps overtime achieve close to 100% automation testing by analyzing where exactly dedicated QA resource is needed and continue optimizing the same. And a formula can be arrived at how much QA resources are needed per product over time.

PERSON DAYS, FIBONACCI AND T-SHIRT METHODS

There are different metrics used to estimate stories.

- Person Days
- Fibonacci Series
- T-Shirt sizing

The goal of estimation should be to get it as accurate as possible so that schedule can be predicted. However estimates are estimates – there should be room left for margin of error. (For approximately 10% margin of error)

Person Days

Person days are the most simplest and straight-forward form of estimation. If Story point is 1, it simply means it takes 1 person day to complete the story.

It is the best option for estimating provided organization/project team have a very thorough estimation tool in place. It takes out lot of abstract method of estimation and makes the entire process more concrete.

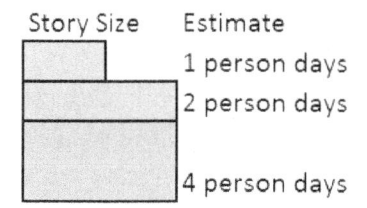

Story Size	Estimate
	1 person days
	2 person days
	4 person days

Story Points	What it means

Example in Person Days	
3	It takes approx. 3 days to complete this story
7	It takes approx. 7 days to complete this story
13	It takes approx. 8 days to complete this story

Fibonacci Series

The next best is Fibonacci, which are the most popular ways of estimating in agile implementation. With Fibonacci numbers, as the complexity of story increases, Fibonacci gives slightly more margin of error in estimation. It is by far the easiest option for most projects.

Before starting to use Fibonacci Series, a reference story is considered – and everything else is estimated relative to that.

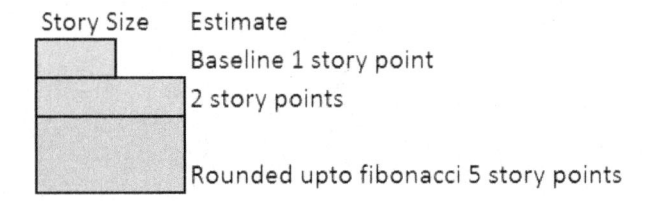

Story Size	Estimate
	Baseline 1 story point
	2 story points
	Rounded upto fibonacci 5 story points

This method is provided with the assumption estimation is a difficult task and using relative story makes it easier to estimate.

Following is an example where 1 story point is mapped to 1 simple story, which intern translates to 1 person day in reference scale.

Reference Story Example:

- Add a simple UI element and backend CRUD. It has taken approx. 1 day to complete. This story is given 1 story point.
- How it is used: Let's say there is a story that has 3 such changes in different services, roughly this is estimated 3 times more complex relative to reference story above, and given 3 story points.

Story Points Example in Fibonacci	What it means
3	It takes 3 story points which roughly maps to 3 days
~~7~~ 8	It takes 8 story points which roughly maps to 6 to 8 days. In Person days example, this was 7 however since Fibonacci series is used, story points 8 is used
13	It takes 13 story points which roughly maps to 13 days

Following are the first few Fibonacci series.

- 1, 2, 3, 5, 8, 13, 21, 34, 55...

From agile usage perspective, this should be good enough. If a story is estimated bigger than 55, then probably all indication is that it is not written well enough to be small.

T-Shirt Sizing

T-Shirt Sizing is another option of relative estimates. It literally is named after T-Shirt sizes (XS, S, M, L, XL etc).

Since it doesn't give the range Fibonacci gives and also since it's not a numerical number it is NOT a good option compared to above 2 options and should NOT be used.

Numerical are easy to measure unit of work and can also help in calculations such as adding up the story points.

Velocity

Velocity is a simple calculation measuring units of work completed in a given timeframe. It helps figure out how much team can achieve in given timeframe. It also helps figure out how team is doing over time compared to itself historically.

Velocity in Agile typically uses Story points as the metric for unit of work.

Sprint Velocity = No of Story points in a Sprint
Release Capacity = Sprint Velocity * No of Dev Sprints

Velocity Chart Example is given in reports section of this book.

Person Days v/s Fibonacci

If there is a matured estimation tool in place, there is no downside to person days based estimate as it is most accurate and it also avoids the additional step of mapping reference story to person days.

Team Sizes are not always constant in practical world. Team size can change (upwards or downwards), personal can change, and skill composition can change. In such scenarios it is much easier to arrive at new velocity using old historical velocity if the estimation uses person days.

Scope To Release

Once story point estimation is complete, team can easily figure out how much can be achieved in a release by multiplying development sprints available with velocity.

Here is an example:

Product Leadership wants to release next version of product in 3 months. Let's say that's around 12 weeks.

Formula	*Example*
Total Sprints Available in Release = Totals Calendar Weeks / Weeks per Sprint	12/3 = ~4 sprints
Total Dev Sprints = Total Sprints – System Testing Sprint	4-1 = 3
Target Story Points per release = Total Dev Sprints * Avg. Sprint Velocity	3*30 = 90
With 10% estimate error margin (min/max)	Minimum Achievable 81 Maximum Achievable 99

Project Team can now target a minimum story points of 81 with high confidence and if things go well then max story points of 99.

Going back to the ordered backlog, two cut lines are drawn – one at 81 guaranteeing product management they will be delivered and they can start planning marketing strategy around it. Another cut line is drawn at approx. 99 as an engineering goal if things to go well what they can achieve beyond 81.

In the diagram below, yellow cut line (cut line 1) shows minimum targeted stories for the release and red cutline (cut line 2) shows best case nice to have scenario.

	SP	Total	
Story 1	13		
Story 2	21		
Story 3	21		
Story 4	13		
Story 5	13	81	cutline1
Story 6	13		
Story 7	8	21	cutline2

Uniform Story Points

Agile teams are self-contained, they know their work items, they come up with their own baseline story, and relatively estimate other stories and execute.

This causes a problem for executive leadership how to compare performance across teams. Let's say in a sprint of 3 weeks, Team A has achieved 50 Story Points, and Team B of same number of resources has achieved 70 story points. This difference could be because of either a) Team B is simply better than A OR b) Team B used a different baseline story point reference mechanism.

While individual teams are not concerned about this difference in velocity as they are only focused on getting better overtime relative to the teams earlier performance, this a real problem for executive leadership since teams do need to be measured for performance for various aspects.

For this reason, organization can try to standardize the story point estimation process across teams to reap benefits of simplifying and standardizing estimation process.

If same story is given to 2 teams, the story needs to be estimated exactly same number of story points. While this has nothing to do with agile development methodology itself, it greatly helps organizations further streamline and automate processes.

This can be achieved with the help of estimation tool custom developed by the organization to suite their products and technology stack.

Teams	Story	Estimation	Story Points
Team A	Develop Login Functionality	Enter inputs in estimation tool to get output in story points	13
Team B	Develop Login Functionality	Enter inputs in estimation tool to get output in story points	13

ESTIMATION TOOL

Organizations can develop their own custom estimation tools to determine the uniform story points.

This also helps in two aspects:

- Helps individual teams estimate more accurately with a baseline reference point.
- Helps executive leadership to compare across teams productivity by using uniform story points across teams. While this doesn't impact the individual project execution itself, executive leadership can look at this and

better measure team performances, deep-dive to understand where needed.

Following is a simple estimation tool example for a layered architecture software application.

Story	UI	Business Logic	Adapter	Total Estimate*
<story 1>	2	4	1	19d

$$*\text{Total Estimate} = (UI*1d)+(BL*3d)+(A*5d)$$
$$= (2*1) + (4*3)+(1*5) = 19d$$

5.3 DESIGN SPRINT

ANALYSIS

A **One Week** design sprint recommended in the hybrid agile model to analyze the release backlog stories, have high level solution architecture defined and mock-ups created and validated. Apart from these CI/CD process and Test Plan also need to be defined.

Analysis involves thorough walk thru of stories by product management with agile team that typically consists of architects, developers and testers.

SOLUTION ARCHITECTURE

Solution architecture defines high-level solution for a given problem. Whether new solution architecture or any changes to existing as a result of release backlog user stories need to be described in the design sprint.

Prototyping using Mock-ups

The most important part of design sprint is to quickly come up with prototype that can be validated with product management and end users.

The mock ups are hard coded, and not much time should be spent on it.

The type of mock ups depend on the interface of the user with the application.

If the application has UI interface, then mock ups are usually hardcoded UI pages without any functionality.

If the application has a non-UI interface, then mock ups need to depict the commands/interface and experience that user would go through using such application.

CI/CD and Source Code

Before development sprint cycles can be kicked off:-

- All the CI/CD (Continuous Integration/Continuous Deployment) processes must be in place with automation enabled to maximum possibility.

This can be achieved outside of development sprints and is a onetime activity for the product.

If it is an existing product, this should be enabled by refactoring what is needed to fully automate CI/CD.

If it is a new product, a skeleton app is created which typically has bare minimum code and can be achieved in short time. Bare minimum code such as a Sample Controller, Model, View with Unit Tests, Extreme Validations Simulator, CI/CD process, Documentation Tools etc.

This involves figuring out source code and branching strategies, automating building of app thru continuous integration and continuous deployment process, setting up running of automation test suites, enabling deploying of the artifact to test environment automatically, running extreme validation/system tests on deployed setup and automating generation of pretty documentation from source code and documentation help files.

This will ensure speed to market from the time the change is introduced to the time it is deployed.

SOURCE CODE AND BRANCHING STRATEGY

Source code management and branching strategy is an extremely important foundational aspect of software development. It determines the success of how smoothly CI/CD process can be built, how smoothly entire process can be automated and how smoothly change can be introduced and managed.

There are various well established branching strategies in use such as development on *"master"*, development on *"develop"* with supporting *"feature"* branches etc.,.

Branching strategy needs to be thoroughly analyzed and adapted prior to development lifecycle kick off based on type of your product.

CI/CD

CI/CD is a process that refers to automated way of achieving Continuous Integration and Continuous Deployment (Continuous Delivery).

It is a set of automated processes that enables following:

- Code Check-ins triggers build automatically.
- Build happens by pulling the dependencies from repo automatically.
- Test Suites are executed automatically.
- Product artifact is built automatically.
- Product artifact is deployed automatically to test setup.
- Extreme Validation and System Tests are run automatically on test setups.

Documentation (in pretty format) is generated automatically from source code and help files.

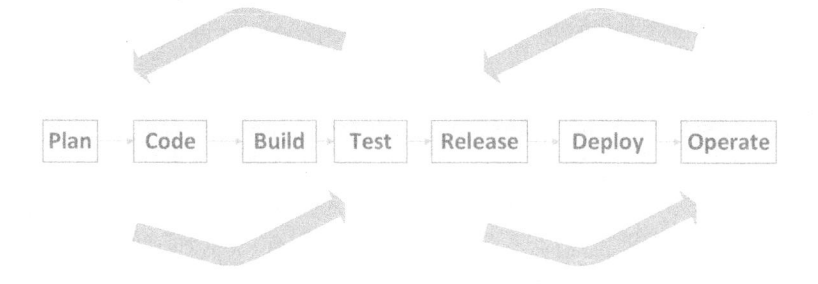

| Plan | Code | Build | Test | Release | Deploy | Operate |

TEST PLAN

During initial sprint, test plan for the project needs to be developed and/or updated with the now available release scope information.

Key elements of Test Plan are:

- Overall test strategy, objectives and processes for the release

- Types of tests to be run and details of same
- Environments used/new to be prepared
- Tools and processes used
- Metrics to be captured

5.4 Dev Sprints and Beta Releases

Sprint

A **sprint** [8] is a short, time-boxed period when a **Scrum** team works to complete a set amount of work.

Sprints lengths are typically 1 week OR 2 weeks OR 3 weeks. Very rarely sprint lengths go beyond 3 weeks. Most widely used sprint lengths are **2 weeks** OR **3 weeks**.

Sprint Planning

Planning sessions

Generally Scrums prescribe having sprint planning session to plan – however these sessions are many times not productive due to several reasons.

Many times story details are not good enough and developer has to go back and analyze a bit to come up with estimate. Estimation process like poker cards take long time, and ineffective as team is estimating on very high level details. It is also very time consuming process.

What can go wrong?

There are few problems with fitting a story into sprint – first, it is impossible to always fit all stories into exactly sprint duration.

For example, if sprint length is 3 weeks (15 working days), since each story complexity is different, some stories could be 8 days, some 11 – in this case it is not possible to squeeze both stories into single sprint (assuming 1 resource), at the same time it is not good use of time to keep waiting for next sprint as well to take up the 2nd story.

If a story gets finished on 11th day, then developer would simply wait for sprint date OR extend current story – and Scrum masters who religiously follow sprint planning/closing will be hesitant to assign new story resulting in resource bandwidth wastage.

At times, developers might get stressed out to forcefully fit in a story to 10 days that lets say takes 12 days.

At times, developers might use the sprint duration as an advantage to simply stretch the story longer and develop slack as it is understood until next sprint nothing meaningful gets assigned.

Recommendations

A simpler and better approach is to assign stories as and when people free up, and as and when previous story gets completed without really getting fixated over sprint planning and closing dates.

However, Sprint planning/closing dates are still used as checkpoints to close a milestone and demo the stories that move to DONE status during that sprint. It doesn't necessarily mean they started in that sprint. Stories that are DONE as of that date are picked up and demonstrated.

An example how this works:

Here is an example of how to use resource bandwidth that opens up in the middle of the sprint, how to tackle stories that don't always fit into sprint length and how to not compromise on architecture and quality.

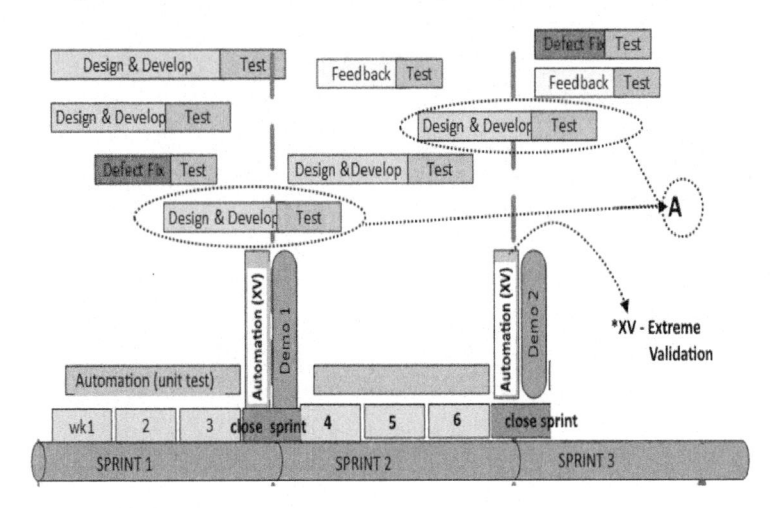

In the diagram above, as (A) is depicting, some stories are not starting and ending on sprint dates.

This process as illustrated in above diagram achieves multiple purposes as follows:

1. Sprint planning and closing happens as usual as per standard Scrum best practices, keeps sprint planning/closing sessions light-weight and milestones are achieved for maximum number of stories.

2. For some Stories, they start and end as per the estimate they fully deserve and as and when resource bandwidth is available – this ensures optimal use of bandwidth, ensures no compromise with architecture and quality.

If stories are assigned thru out, does that break time-boxing and negate time-boxing advantages?

Time-Boxing is an extremely important concept as explained early in the book.

It is important to understand, the artifact deliverable in agile process is "Story". There is no such thing as delivering a sprint.

Sprint outlines overall time-boxing for group of stories. But by assigning stories thru out as and when stories get done, it does NOT break any time-boxing rules because the "Story" still has its own time-boxing – the only change being, it now need not be squeezed into a very strict milestone if it comes at the cost of compromising architecture and quality.

This is slightly similar to Kanban process however, not exactly as it is still achieved within the overall Scrum process and the story will have estimation and its own time-boxing. Also, if the sprint length is correctly identified for the type of product/application, then such stories spilling over to next sprint will be very small quantity.

If stories are assigned thru out, how to track sprint velocity?

One of the problem with above approach is track what was achieved in the sprint if the story got assigned midway and didn't get completed in the same sprint – however, this doesn't make much difference to productivity as such – it only impacts measurement of productivity.

Following method can be used to solve this issue.

There is one option should the story be split in that case, even if it's not testable. However, one should try to keep things simple - and if this option to be followed then it messes up lot of things such as what if QA team needs to refer to the story in its entirety.

A simpler option if tool allows this support (or hopefully in future) is to allocate the stories to multi-sprints and split story points to both sprints.

Most agile tools today don't allow this feature, in which case you can add a custom field "Story Point division" and manually enter the story points in different field for each sprint and in the charts tool, this field need to be used to generate velocity charts correctly.

Velocity can also be calculated by using the release velocity and calculating the average per sprint. In this case, above options are not needed.

STORIES AND SUB-TASKS

Story development typically consists of design, development, error handling and unit testing of of one or more logical code portions, writing test-scripts, extreme validation (simulator) testing, documentation, demo and acceptance.

Each of these can be tracked as sub-task within the story. It also allows assigning multiple resources to a story as most agile tools allows assigning resources at Sub-Task level.

However a sub-task must be avoided to be assigned to more than one person as the accountability becomes very much muddied in that situation. In such situation separate sub-tasks needs to be created.

Sub-tasks also have their own, but simplified workflow – New, In-Progress, Done. Sub-tasks should not be confused to stories, hence the workflow status is only limited to that step. They are there to help better tracking of various steps in a story.

Example sub-tasks for a story is given below:

Sub-Task	Status
Design	DONE
Development	DONE
Unit Testing	DONE
Writing Test Scripts	DONE
Extreme Validation (Simulator) Preparation	IN-PROGRESS
Extreme Validation Testing	IN-PROGRESS
Documentation	IN-PROGRESS
Demo	NEW

STANDUPS

Whether Waterfall OR agile, frequent checkpoints are needed to check status on the project, discuss dependencies, blockers and other issues. Agile Scrum Methodology introduces term "Standups" for typically daily stand ups to discuss team progress – what was achieved, what is planned and blocker issues.

What can go wrong?

It is always good to have frequent touch points for better coordination among the team and remove any blockers that warrants a meeting. However, these should not be followed religiously as per Scrum guidelines. Some Scrum masters literally say a stand up must be daily, exactly 15 mins. Why 15 mins – why not 14, 16?

15 min daily standups also gives a feeling to developer – let's get thru these 15 mins somehow – and productivity dips once the meeting is over.

Software development is a creative work – often times one needs dedicated time to preserve chain of thought and come up with good design. Daily checkpoints can be pretty disruptive to this work.

What works?

2 touch points a week for around 30 mins each works best in author's experience. It gives enough continuity to developers not to get their work short circuited with daily standups.

KEY METRICS LOGGING

While it is important to reduce the "standup" kind of checkpoints for entire team, at the same time it also important to establish a process to do the daily logging of key metrics and logging of data when workflow progresses for stories and defects.

These metrics will then feed into various project management reports and results in highly transparent environment.

The metrics also help automate the processes to maximum extent.

For example if a story is moved to "QA Ready" by developer, the tester assigned will immediately know it's available for testing and will pick it up without waiting for next stand up.

If a story is blocked, by blocking the story in the tool, the Scrum master will immediately come to know that story is blocked and takes necessary action without losing valuable time.

For optimized process all such reports should be delivered thru automation e-mail subscriptions, as they get delivered right into your e-mail inbox.

More reports are explained in Part 6 of this book.

What to log and when?

Following are some of key metrics developers/testers/team should log for stories and defects.

What to log	How frequently
Story Points	Only log once when story is assigned and started. Do not change original story points irrespective of if time tracking estimate turns out to be more or less than original story points.
Story Status	Update when workflow step changes
Sub-task Status	Update when Sub-task status changes
Comments	Log as and when an important update is there, at a minimum once a week
Time Tracking	Log time in days or hours at a minimum once a week OR when story/sub-task is done.
Blocker flag	Update as and when story is blocked/unblocked
Defect Status	Same as story status, log when defect workflow step changes

DESIGN & DEVELOPMENT

Once a feature is taken up for implementation, it goes thru typical SDLC of design, development, unit testing, functional testing and system testing phases.

Often times this design ends being very crappy due to:

- Scrum masters strict "sprint" process pressurize the developers.
- Developers who just want to get thru the Scrum do work thru short cuts to get thru day's work.
- Sometimes senior management may encourage such "short-cuts" behaviors for various reasons such assumptions that it may bring speed to market and early market share.

Agile by itself doesn't guarantee success if fundamentals of building software are not followed.

Error Handling and Error Codes

Any good software design needs to incorporate proper error handling and error codes. Often in agile project management, architecture principles take a hard hit due to various reasons.

Every story must have a section – what are possible errors and how they are caught and reported.

It is neither easily possible nor it is cost effective for companies to recreate every scenario that customers faces and have every possible customer setup in their test bed.

If a customer issue is reported, therefore it is important that software application has the ability to detect what exactly happened with the help of error code/or logs without the need to recreate the scenario.

Every story must have a check as part of acceptance criteria if error codes are thought thru for all the scenarios and coded accordingly.

Here is a simple example of how error handling can help –

```
try {
        //your story logic
        //application will throw business exception upon error
        condition encountered

        errroCode = new ErrorCode('BE01','The following business
        exception has occurs – <exact scenario here>');
        throw new BusinessException(errorCode);
        Logger.info("An error occurred" + errorCode.getCode());

} catch (Exception e) {
        errroCode = new ErrorCode('SE01','The following business
        exception has occurs – <exact scenario here>')
        throw new SystemException(errorCode);
        Logger.error("An error occurred" + errorCode.getCode());


}
```

This is just a pseudocode, in the above example application throws Business, or System Exception with exact error code and same is logged.

With the error code thrown, the controller layer can determine what to show to the caller system or end user.

For example, view layer can display "BE01 An error occurred."

And by looking at the logs, monitoring applications and support teams can easily determine what exactly happened. In this scenario, there is no reason to recreate the issue to analyze the error.

Monitoring applications can go one step further and automatically analyze the error and automatically respond back to the user with corrective action.

AUTOMATION TEST SUITES

Unit Testing

Unit testing is a testing methodology where individual units of source code is tested with stubbed data. Unit tests allow you to make big changes quickly. Unit Tests helps complete coding faster and merge faster as you have verified code is working as expected. This is one of the bare minimum things needed for any agile projects to establish continuous integration and continuous deployment (CI/CD) automation process.

Extreme Validation Testing

Unit testing refers to testing individual units of code.

For larger integration testing traditionally QA team performs end to end system tests using test infrastructure that is production like. However, it is not practical to prepare every single permutation and combination of production like systems.

The focus of extreme validation methodology adopted is to maximum simulate the system behavior so that expensive testing methods using real world physical prototype can be bare minimum.

In a nutshell, extreme validation is -

- More tests using simulations
- Less focus on physical prototype

It is a common practice for NASA astronauts to spend time in simulators to undergo various possible permutations and combination scenarios.

Similarly a software application should have a robust way of simulating actual system behavior it is interfacing with.

The purpose of such a simulator product is to mimic the actual production systems that application is interacting with.

It can be developed using various technologies depending on the technology stack of your application. The simulator product will output mocked test data that is close to production system.

Functional Testing

Functional Testing is same as System testing except it is focusing on particular feature being tested. Products can adapt functional testing using real integrated systems or simulators as described in extreme validation section.

System Testing

System testing refers to end to end application testing using production like systems. System testing validates the complete full integrated software application.

Regression Testing

Regression testing refers to running full suite of all available test cases – this includes unit testing, functional testing, extreme validation testing and system testing. This is run to ensure test suite is constantly updated as change is introduced and nothing is broken in the system as and when change is introduced to production software.

The relative cost of fixing a defect raises exponentially as the project progresses. Thorough testing is not only good for company reputation and customer satisfaction, it is also needed to sustain the product with minimal cost.

Test Scripts

Test scripts are written for each story in short and concise manner with information - input steps, expected results and actual result.

These are best maintained in tool that supports updating test results and running multiple cycles of tests.

Story Workflow

A story progresses from one step to another thru various states. Following is a sample workflow for story. Most agile tools support customizing workflow as per your needs.

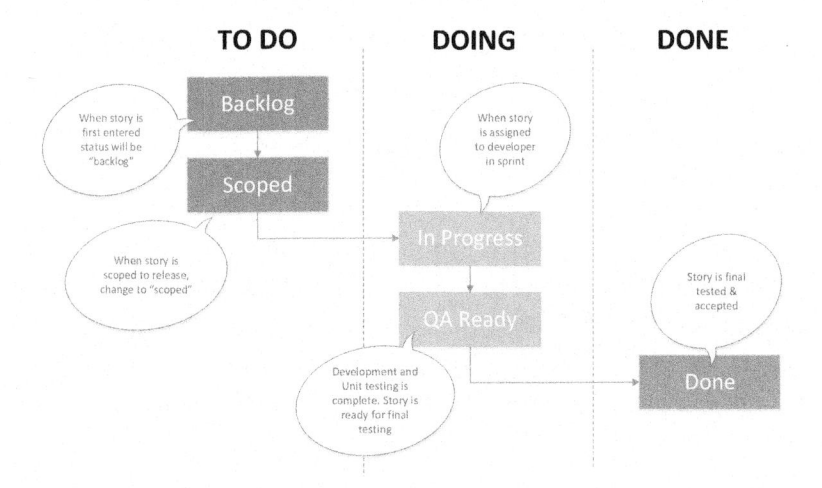

TO DO — **DOING** — **DONE**

- Backlog: When story is first entered status will be "backlog"
- Scoped: When story is scoped to release, change to "scoped"
- In Progress: When story is assigned to developer in sprint
- QA Ready: Development and Unit testing is complete. Story is ready for final testing
- Done: Story is final tested & accepted

Backlog	Story status when story is entered in product backlog.
Scoped	Story status when story is scoped to a release.
In Progress	Story status when story is taken up for development.
QA Ready	Story status when story development and unit testing is complete. Ready for QA Testing.
Done	Story status when story is final tested and accepted.

DEFECT WORKFLOW

Once the story reaches "QA Ready" stage, any issues found in story after that would be logged as defect.

Do defects have story points?

No. Defects should NOT be given any story points or estimated as they add no productivity in the end to product development. Defects should be given **zero** story points.

How does defects planning work in sprint cycle?

Defects are assigned like any other work item from FLAT list of items. But there are few key difference w.r.t to story assignments.

- Defects are NOT given any story points nor are they estimated.
- Generally the developer who originally worked on portion of story should be the owner of defects. This drives more accountability and competitive spirit within the team to produce high quality code.

- Defects are picked up proactively by the developer without any formal sprint planning sessions.

Defects usually have workflow as given below

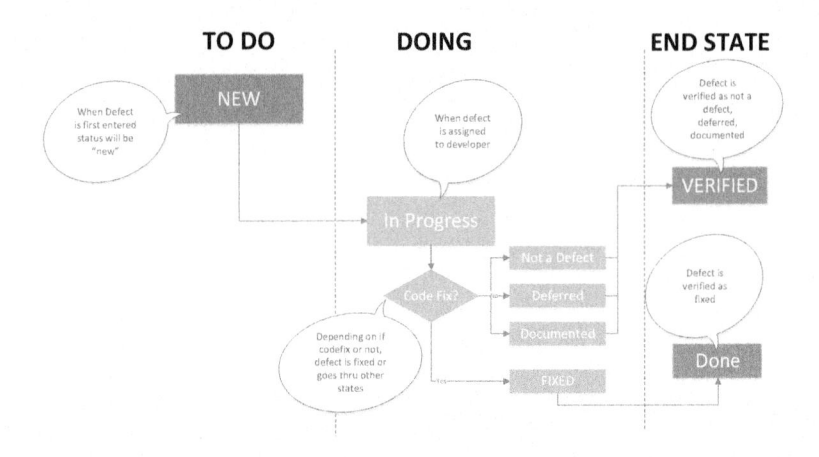

TO DO   DOING   END STATE

(Workflow diagram: When Defect is first entered status will be "new" → NEW → In Progress (When defect is assigned to developer) → Code Fix? → Not a Defect / Deferred / Documented / FIXED. Depending on if codefix or not, defect is fixed or goes thru other states. Not a Defect, Deferred, Documented → VERIFIED (Defect is verified as not a defect, deferred, documented). FIXED → Done (Defect is verified as fixed).)

New	Defect status when defect is open
In Progress	Defect status when defect is taken up for fixing
Fixed	Defect status when defect is fixed and unit tested.
Not a Defect OR Deferred OR Documented	Defect status when developer determines defect will not be fixed for one of the reasons whether it is not a defect, or defect to be deferred/moved, or documented
Verified	This is end state if defect is not fixed. Reporter accepts developer assessment
Done	This is end state if defect is fixed. Reporter tests and accepts the fix.

API & USER DOCUMENTATION

Each story needs to be documented as it needs to be consumed by other systems as an API and/or end user.

Documentation tools allows one to extract documentation from source code and read me files to generate pretty HTML files, therefore – this step should be maximum automated to generate HTML/PDF files. Documentation should be done as and when story are done as oppose to keeping it as a separate step at the end of the project.

SPRINT DEMO

At the end of the sprint, stories completed in that sprint are demonstrated to all stake holders. A recording of the same should be made and uploaded to easily accessible storage.

An improvised option here is to record the story demos as and when they are done during the sprint and share it with stake holders.

For example let's say, it's a 3 week sprint and at the end of 1 and ½ week one story is completed while other stories are still in progress, then that story need not wait till end of sprint for it to be shared with stake holders. This should be an offline process to not to cause any meeting overheads.

This process should be automated – a demo link is tagged to the story when the workflow moves to DONE.

FEEDBACK LOOP

Feedback is collected and tracked as separate story. This helps continuous feedback loop and early validation.

Does feedbacks make project scope increase?

Due to continuous engagement thru out the project, including engaging product manager at early mockup decisions, typically any feedback in a well-executed project will be minimal and can be easily accommodated within the usual estimate buffer. An additional some time can be allocated at the planning time to accommodate for feedback.

SPLITTING STORIES

In a multi-sprint release, sprints are there to ensure certain milestones are met. A story need not start on planning day and finish on closing day. Though it is ideal, it is not practical and often leads to inefficiency if such a hard rule is made.

Since release consists of multiple sprints, this is nothing to be too concerned about. It is impossible to split the stories such that they all add up to exactly the no of days available in the sprint. Stories are best executed as and when people free up from flat list.

In this case one the following 2 options should be followed --

Is the completed portion of the story "independent, testable, demo-able, consumable"?

If yes, then story can be split with suffix such as part1/part2 and completed portion can be closed.

If no, then the story needs to be simply moved to next sprint. It will skew the velocity calculations as most agile tools don't allow assigning one story to multiple sprints and divide story points, however this is still a simpler and better option than to forcefully split the story if completed portion doesn't satisfy "independent, testable, demo-able, consumable" criteria.

Sprint Closing

This marks the sprint closing, and the sprint can be closed in agile tools.

Beta Release

To reap the full benefits of agile process, the product should be released as frequently as possible.

A Beta release is recommended at the end of each sprint, to subset of customers to validate the features completed. Such strategy is also called "pilot release".

5.5 Release Candidate Sprint

Release Candidate

At this stage the product has gone thru several beta releases to subset of customers and product management. All the feedback is collated, more features are rolled out each sprint, defects are fixed – now the product is ready for general availability to entire customer base.

Prior to making it general availability a sprint called "Release Candidate" sprint is extremely useful to address:

- Code branching and Release tagging
- Final code fixes
- Full regression suites
- Non-functional testing

A small dedicated sprint is allocated to do these activities. The sprint length itself can be adjusted based on complexity of the project, if there is any need to shorten this length OR make it larger duration compared to development sprints length.

Non-Functional Testing

Non-functional testing [7] is the testing of software application for non-functional requirements such as browser compatibility, device compatibility testing, load testing, scale testing, continuous hours of operation testing etc.,,

Usually non-functional testing is not performed during regular sprint testing unless it is fully automated. While it can be performed, since the product is still going thru beta releases it is a tradeoff cost v/s benefit. For this reason, RC Sprint can be used to perform this testing.

5.6 Release To Production (GA Release)

Product has already been rolled to subset of internal and/or external customers each sprint, with each beta release having incremental feature set.

For some organizations this could be simply releasing the product only to internal product management for acceptance testing. However, where possible to reap full benefits of agile methodology it should be release to subset of customers.

Now the product is at a stage to be released and made generally available. Generally Available means all the product strategies such as pricing, marketing, support models etc are in place and product is tested for scale and load and made available to all customers.

Checklist

Checklist is used as final check prior to releasing product to production.

Since processes are automated to maximum extent, ideally these checklist should be auto generated as much as possible without much manual effort.

DEPLOYMENT

The final step of release is to deploy the code to production – depending on if a product type (Downloadable, Saas, etc) deployment process is followed.

Once the release is complete, project team can get together and do a lessons learned session on how to improve release over release. These retrospective /lessons-learned sessions are best to happen once per release as oppose to once per sprint to minimize the unnecessary overhead.

Agile tools get cluttered very quickly. It is important to close the release in agile tools as well.

PART 6: **HYBRID AGILE** PROJECT REPORTS

6.1 REPORTS OVERVIEW

There are various stake holders in a project – and agile project reports becomes very essential in keeping project on track and ensuring timely action is taken for project success.

Following are some of the most useful reports for any software product development project.

Report	Purpose	Audience
Release Report	This would provide details around what stories are being implemented as part of release and what is the status of same.	All stakeholders
Sprint Report	This report will track each story taken up in the sprint and what is the status of same.	Project Lead and Engineering Management
Time log Report	Any successful project depends on its people. It is essential to optimally utilize people and their skills. This reports give details of time logged by each project member so that engineering manager can make necessary adjustments as needed.	Engineering Manager
Velocity	This is a very popular	Engineering

Report	report to measure team productivity over time. It also gives a good view of productivity trends over time using which organizations can improvise on various aspects.	Management
Product Backlog Report	All successful organizations have good roadmaps – This report gives a view into product vision in future.	All stakeholders

6.2 Release Report

Release report consists of graphical representation of release burndown to provide data around if release is on track/behind track or ahead of track.

In the below example, release was targeting 144 points at a minimum and 176 was maximum possibility to the best of estimate. The burn down diagram shows, how many story points are achieved (burned) each sprint and how many remaining.

Release Burn Down By Sprint

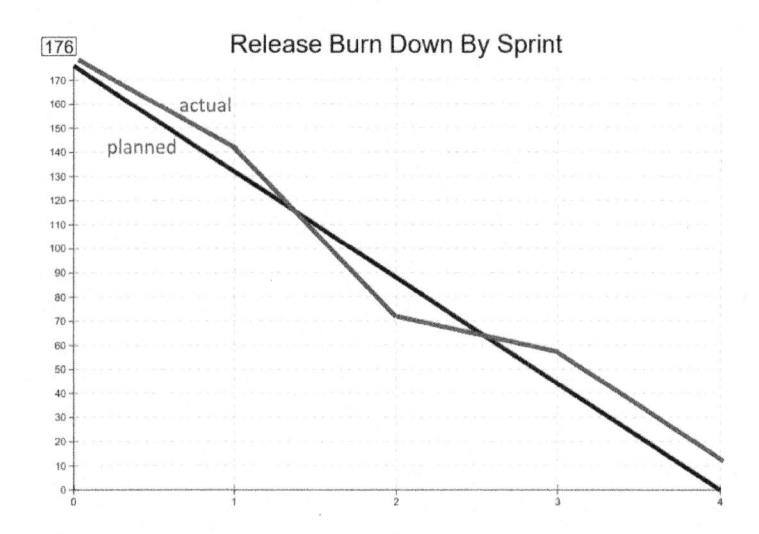

[176]

Along with the burndown chart, a report of flat list of done v/s remaining stories are useful update to stakeholders.

Done	SP	Remaining	SP
✓ Story 1 <title>	13	Story 5 <title>	8
✓ Story 2 <title>	8	Story 6 <title>	8
✓ Story 3 <title>	13		
✓ Story 4 <title>	5		

6.3 SPRINT REPORT

Similar to release report, sprint report provides insight into what is achieved in a sprint. A sprint burndown looks very similar to release burndown, however sprint by itself is so small – a sprint burndown report may not be very useful.

Sprint Report When Sprint is In-Flight:

Primary purpose of sprint report when sprint is in-flight is to be on top of things and execute the sprint smoothly. This is the main report for Scrum Master and Technical Lead to manage day-to-day work.

Few key things that are useful are – for each story, who it is assigned to, is it ready to be tested/demoed, and are there any blockers, what are the daily updates, what are the sub-tasks etc. so that appropriate action can be taken in timely manner.

Story	Assigned To	Status	SP	Blocked?	Sub-Tasks	Update Comments
Story 1 <title>	Steve	Done	13	No	a) ST1-Done b) ST2-Done	
Story 2 <title>	Bob	QA Ready	8	No		
Story 3 <title>	Kumar	In Progress	13	Yes		Opened a ticket with vendor
Story 4 <title>		Scoped	5	No		

Sprint Report When Sprint is DONE:

A simple report that shows how many stories are DONE, how many story points are achieved, Quality Metrics, and optionally how hours logged are useful sprint data.

Sprint Completion Summary

Sprint No: Q1 S1

Stories		Defects		Feedback		Points	
Completed	3	Fixed	3	Received	2	Story Points	30
Moved	1	New	2	Completed	1	Hours Logged	280

Sprint Details

Story/Defects	Type	Status	Story points
Story 1 <title>	Story	Done	13
Story 2 <title>	Story	Done	8
Story 3 <title>	Story	Done	13
Story 4 <title>	Story	Done	5
Defect <1>	Defect	Fixed	0
Feedback Story <1>	Story	Done	3

6.4 VELOCITY REPORT

Velocity is a measurement how much a team can achieve within a sprint. This measurement is typically done in story points.

It can primarily help leadership to

- ✓ Reasonably forecast future (what can be achieved within a timeframe).

- ✓ It can help understand aggregate velocity of team.
- ✓ It can help diagnose any problems and optimization opportunity for productivity and efficiency.
- ✓ If unified story points estimation method is followed by organizations, then it can also provide meaningful comparison across other teams.

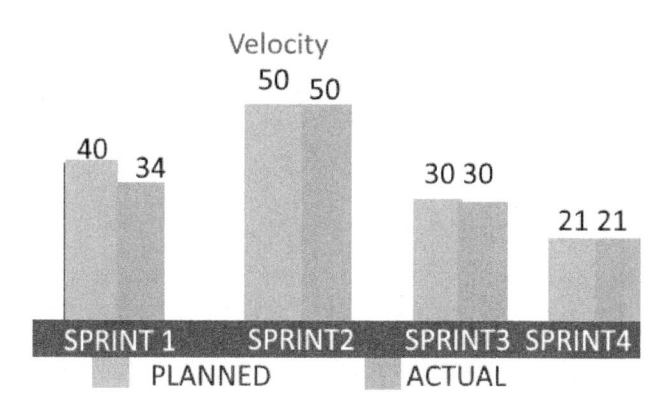

Velocity

	SPRINT 1	SPRINT2	SPRINT3	SPRINT4
PLANNED	40	50	30	21
ACTUAL	34	50	30	21

6.5 Time Log Report

In a project team not every member is operating at the optimal efficiency, skill, motivational and accountability levels.

For engineering managers and senior management, it is important to have visibility to who is working on what, and on how exactly the time is spent.

Time logging helps spot manual processes that can be automated, help identify employees that are performing above, at par or below expectations and act on same, cultivate discipline in the system and can also contribute to peer motivation.

Time logging on story can be done as frequently as convenient but at a minimum once a week.

People managers often measure employees through various WHAT and HOW aspects of the employee. Time logging reports are only one small part of it, however they can be useful to analyze and provide timely feedback to employees as needed. Such reports can be analyzed once a sprint.

Example time logging report is given below

Sprint Duration: Feb 24th to March 7th

Resource	Project	Story	Days Logged
Steve	Project A	Story SU01	8d
Bob	Project A	Story SU02	5d
	Project B	Story KY01	4d
Kumar	Project A	Story SU03	8d

6.6 PRODUCT BACKLOG REPORT

Product Backlog reports are simply all the OPEN backlog stories that are NOT in any release and not in DONE state yet, ordered by priority ranking. This will help executive leadership to keep a watch on backlog health.

Sample backlog report template:

Story #	Title	Status	Label	Story Points	ROI	Priority#
SU01		Backlog			$	1
SU02		Backlog			$	2

PART 7: HYBRID AGILE AREAS OF EXCELLENCE HANDBOOK

To recap what book has covered so far -- while practicing agile Scrum methodology it is important to make sure, the problems it causes as specified in Part 2 are solved by customizing and optimizing the methodology.

Below section revisits these goals and captures what these goals are, what can go wrong and what are some of the customizations and optimizations that can be done to address the same and achieve these goals.

7.1 EXCELLENCE IN PRODUCT STRATEGY

What is it?	Product Strategy is the strategic vision of the product. It defines the product roadmap, who are the product customers, who are the customers of product, what value it adds to customers, how it's priced and distributed.
What can go wrong?	Tactical product features are those that are done primarily to meet certain urgent needs such as capturing market share. If it is being done to ultimately reach the destination of strategic vision, then it is OK. However, it should be discouraged if such practice has taken roots with belief "we are agile, we can adapt quickly later".
	Constant change of product strategy not driven by market changes but rather driven by lack of vision (as *"the process allows it"*)

	may lead into "herding cats" kind of situation – leading to many uncontrollable entities in product evolution.
What best practices to follow?	Product management should be trained about the high-maintenance costs of doing tactical solutions more frequently than necessary. While tactical solutions are absolutely necessary at times, it should not be misused as a standard operating procedure. Tactical strategy is a subset of product strategy and vision – it should not become equal or superset of strategic vision.

7.2 Excellence in Customer View

What is it?	Customer View is the customer priority and usage pattern of the product.
What can go wrong?	One might build a great product, but if customer's priority and usage patterns of the product are different, the product will have pretty big gap in-terms of its purpose.
What best practices to follow?	A solid beta/pilot strategy must be established to rollout products to subset of customers early to get customer view on the product throughout project lifecycle. Beta/Pilot strategy also allows identify critical gaps in the product before going to wider audience. This will mitigate risks such as bad branding to the company and

product due to an unforeseen quality or other issue in the product.

7.3 Excellence in Architecture

What is it?	Software Architecture refers to the fundamental structures and frameworks of the software application built by adhering to few must-have principles such as reusability, modularity, configuration driven, service oriented and many other depending on type of application and software stack. It functions as the blue print of the system.
What can go wrong?	Architecture principles must not be cut short with agile process as an excuse. For example hardcoding what should be configuration driven due to time pressure, not thinking thru modularity due to time pressures, not coding properly for extensibility due to lack of vision etc.,
What best practices to follow?	A solid architecture practice must be mandated by customizing Scrum process as described in Part 4 and other parts of this book as necessary.

7.4 Excellence in Quality

What is it?	Software Quality refers to how well the product is working as designed (in

	functional and non-functional aspects) without defects and adhering to user requirements.
What can go wrong?	Not enforcing process around maximum automation with unit testing, not having simulator design patterns in place for extreme validation, poor sprint planning and time pressures can lead to lack of quality in the product.
What best practices to follow?	A solid unit testing and extreme validation methods are mandatory to project quality success. Extreme Validation requires a robust simulator product developed to help achieve testing of 100% permutations/combinations of production like system without the need to brunt the cost involved in coming up with every combination of production systems. Simulation product also allows to do automation of extreme validation without actual production backend systems in place.

7.5 EXCELLENCE IN INNOVATION

What is it?	Innovation refers to original and more effective way of solving a problem or identifying an opportunity and solve.
What can go wrong?	With task assignments as given by Scrum Master and Management and strict sprint

	guidelines, it can lead to a situation where an otherwise innovative and creative developer might soon be stuck in a routine and stressful environment only focusing on tasks as given.
What best practices to follow?	There are various ways innovation culture is established in an organization such as solid idea collection, prototyping, and validation of prototypes.
	One of the factors is to ensure Scrums are run by experienced Scrum masters. Scrum Masters / Project Leads should have technical knowledge and leadership skills to get to the point as quickly as possible, ability to provide solutions/resolve issues intelligently. This enables development teams to focus less on getting caught in inefficient loops and focus more on story delivery and innovation.
	Innovation can happen within the same story or it can happen by coming up with brand new story or idea.
	Another factor is to have story type "Prototype Story" in the standard project lifecycle itself and during release estimation process certain percentage of bandwidth is allocated to allow development team to come up with "Prototype" type stories and develop and demonstrate the same.

7.6 Excellence in Automation

What is it?	"Location, Location, Location" is the well-known mantra of good real estate investment strategy. Similarly, the mantra of Agile Methodology should be "Automation, Automation, Automation". Automation refers to functioning of processes and application functions with minimal human inputs.
What can go wrong?	Inexperience in the team composition and skillset problems can lead to lack of automation.
What best practices to follow?	All the processes and application functions with tools need to be fully automated to the maximum possible extent reap the benefits of agile methodology. These include automating from planning stage to the deployment stage at each and every step. Automation includes both product related such as builds of artifact, unit testing, extreme validation, CI/CD process and also processes such as managing story workflows, generating and managing documentation etc. Various e-mail reports can be triggered, delivered right into your e-mail inbox, to further optimize these processes and track information in timely manner. Some of the key report examples are explained in Part 6.

7.7 EXCELLENCE IN SPEED-TO-MARKET

What is it?	This refers to how quickly one can take the product to market and how quickly one can release incremental change. It is one of the foundational goals of agile methodology.
What can go wrong?	Lack of well-established CI/CD process and lack of skillset and experience in team can lead to not achieving ideal speed-to-market.
What best practices to follow?	This is primarily achieved by a very solid CI/CD process (and required talent hired) and a very well established beta/pilot strategy to customers.

Part 8: Scaled Agile Considerations

So far we have looked at how to customize and optimize Agile Scrum Methodology and apply for single agile team product development.

However, often in real world, in large organizations multiple agile teams are developing multiple products – that is okay! But if these products are inter-connected and consuming each other, then the agile teams need lot of coordination to ensure each of these products are released during development cycles and system tested.

Scaled Agile Framework [12] is introduced to solve this problem primarily. This book doesn't go in detail into Scaled Agile Framework (SAFe) implementation.

However, the essence of agile Scrum development remains the same whether one product or multiple inter-connected products except in scaled environment the product requirement co-ordinations, program timings, coordination and governing related aspects come into picture.

For example, "Release Candidate" is released at the end of development sprints in single-product development, but in scaled multi-agile team development, early releases are made, called as "Program Increments".

For example if there are four development sprints, a Program Increment is released at end of second sprint for both Product A (agile team A) and Product B (agile team B) so that early system testing can be performed.

Revisiting the 7-step Hybrid Agile Scrum lifecycle given in section 4.1, now 2 additional steps are added at a high-level as illustrated below.

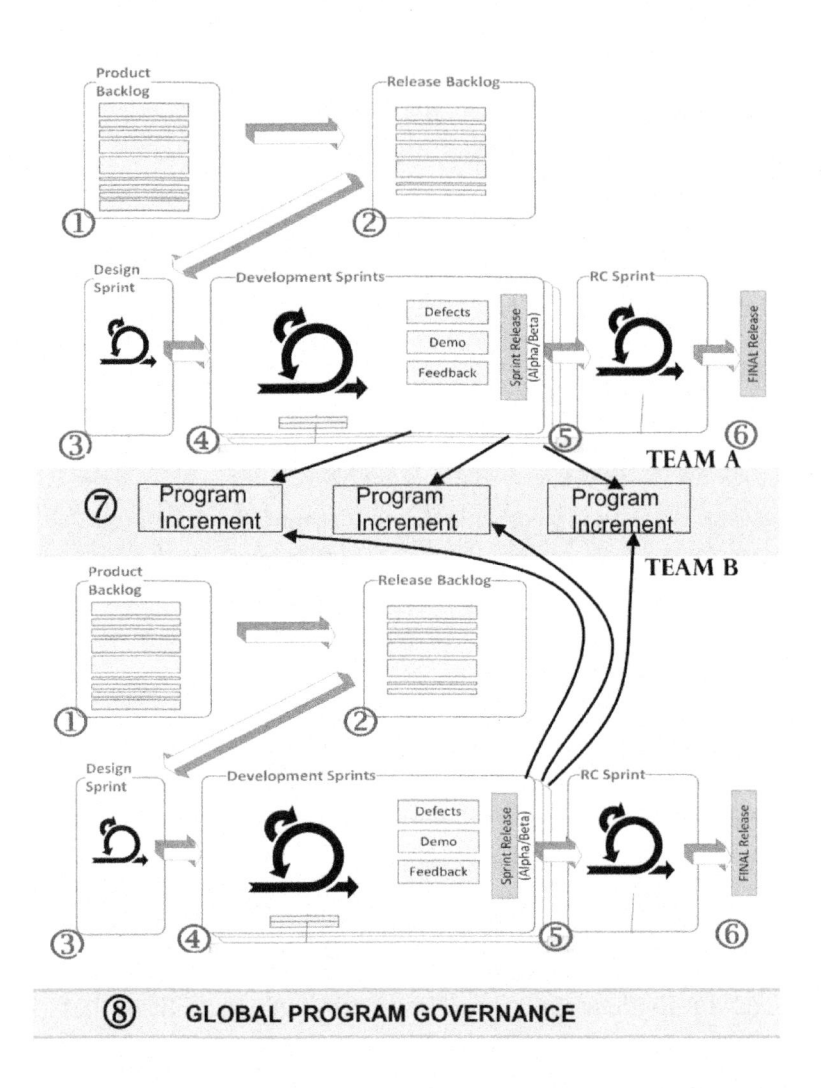

7. **Program Increments** are delivered by each product at pre-determined milestones (For example, in below, a Program Increment is delivered at the end of Sprint 2. – Program Increment is similar to Release Candidate build, except product is still going thru development sprints.

8. **Global Program Governance** is established for the projects to handle prioritization between projects, coordination, dependencies, issue resolutions, testing coordination and other.

This PI Build goes thru system tests that otherwise would have happened only at the end of development sprints.

About The Author

Vijayakrishna R N (Vijay) has rich experience in Fortune 25 /Fortune 200 multinational companies at USA and India as Senior Developer, Solution Architect, and Engineering Manager.

He has 20+ years of experience in building cutting edge Software Applications in various domains and software stacks.

Vijayakrishna R N is a Bachelors of Engineering (**BE**) graduate from **R.V. college of Engineering**, Bengaluru with post grad **MS** in Software Engineering from **Brandeis University**, Waltham, MA, USA.

Author E-Mail: Vijayakrishna.RN@gmail.com

Book blog: www.hybridagile.org

REFERENCES

[1] https://www.bls.gov/bdm/us_age_naics_00_table7.txt
[2] https://www.investopedia.com/financial-edge/1010/top-6-reasons-new-businesses-fail.aspx
[3] https://agilemanifesto.org/
[4] https://en.wikipedia.org/wiki/Agile_software_development#Agile_software_development_practices
[5] https://www.Scrum.org/
[6] https://en.wikipedia.org/wiki/INVEST_(mnemonic)
[7] https://en.wikipedia.org/wiki/Non-functional_testing
[8] https://www.atlassian.com/agile/Scrum/sprints
[9] https://en.wikipedia.org/wiki/Responsibility_assignment_matrix
[10] https://hbr.org/2018/12/how-timeboxing-works-and-why-it-will-make-you-more-productive
[11] https://en.wikipedia.org/wiki/User_story
[12] https://www.scaledagileframework.com/

DISCLAIMER

Although the author have made every effort to ensure that the information in this book was correct at press time and while this publication is designed to provide accurate information in regard to the subject matter covered, the author assume no responsibility for errors, inaccuracies, omissions, or any other inconsistencies herein and hereby disclaim any liability to any party for any loss, damage, or disruption caused by errors or omissions, whether such errors or omissions result from negligence, accident, or any other cause.

ONE LAST THING...

If you enjoyed this book or found it useful I'd be very grateful if you'd post a short review on Amazon. Your support really does make a difference and I read all the reviews personally so I can get your feedback and make this book even better.

Printed in Great Britain
by Amazon